PUBLISHER COMMENTARY

We print NASA's handbooks and standards for the convenience of those that use them on a daily basis. We print all of these a full 8 ½ by 11 with large text so they are easy to read. Yes, color books are expensive to print so unless the information relies on the use of color for proper interpretation or understanding, we print most books in black and white to keep the cost down. All these documents are available for download for free from NASA, however printing them all over a network printer would take days.

Why buy a book you can download free? We print this so you don't have to.

All these books are available for free download from the government web site. Some are available only in electronic media. Some online docs are missing pages or barely legible.

We at 4th Watch Publishing are former government employees, so we know how government employees actually use the standards. When a new standard is released, an engineer prints it out, punches holes and puts it in a 3-ring binder. While this is not a big deal for a 5 or 10-page document, many NIST documents are over 100 pages and printing a large document is a time-consuming effort. So, an engineer that's paid $75 an hour is spending hours simply printing out the tools needed to do the job. That's time that could be better spent doing engineering. We publish these documents so engineers can focus on what they were hired to do – engineering. It's much more cost-effective to just order the latest version from Amazon.com

If there is a standard you would like published, let us know. Our web site is www.usgovpub.com

www.usgovpub.com

Copyright © 2019 4th Watch Publishing Co. All Rights Reserved

List of Other NASA Publications Available on Amazon.com:

NASA-STD-5001B	Structural Design and Test Factors of Safety for Spaceflight Hardware
NASA-STD-5006A	General Welding Requirements for Aerospace Materials
NASA-STD-5008B	Protective Coating of Carbon Steel, Stainless Steel, and Aluminum on Launch Structures, Facilities, and Ground Support Equipment
NASA-STD-5009A	Nondestructive Evaluation Requirements for Fracture-Critical Metallic Components
NASA-STD-5012B	Strength and Life Assessment Requirements for Liquid-Fueled Space Propulsion System Engines
NASA-STD-5019A	Fracture Control Requirements for Spaceflight Hardware
NASA-STD-5005D	Standard for The Design and Fabrication of Ground Support Equipment
NASA-HDBK-8739.21	Workmanship Manual for Electrostatic Discharge Control
NASA-HDBK 8739.23A	NASA Complex Electronics Handbook for Assurance Professionals (Color)
NASA-HDBK-8719.14	Handbook for Limiting Orbital Debris (Color)
NASA-HDBK-8709.22	Safety and Mission Assurance Acronyms, Abbreviations, and Definitions
NASA-HDBK-7009	NASA Handbook for Models and Simulations: An Implementation Guide For NASA-STD-7009 (Color)
NASA-HDBK-8739.19-2	Measuring and Test Equipment Specifications NASA Measurement Quality Assurance Handbook – Annex 2
NASA-HDBK-8739.19-3	Measurement Uncertainty Analysis Principles and Methods NASA Measurement Quality Assurance Handbook – Annex 3
NASA-HDBK-8739.19-4	Estimation and Evaluation of Measurement Decision Risk NASA Measurement Quality Assurance Handbook – Annex 4
NASA RCM	Reliability-Centered Maintenance Guide for Facilities and Collateral Equipment

www.usgovpub.com

Copyright © 2019 4th Watch Publishing Co. All Rights Reserved

	METRIC/SI (ENGLISH)
NASA TECHNICAL STANDARD	**NASA-STD-5009A**
Office of the NASA Chief Engineer	Approved: 2018-06-19 Superseding NASA-STD-5009 (Baseline)

NONDESTRUCTIVE EVALUATION REQUIREMENTS FOR FRACTURE-CRITICAL METALLIC COMPONENTS

APPROVED FOR PUBLIC RELEASE – DISTRIBUTION IS UNLIMITED

NASA-STD-5009A

DOCUMENT HISTORY LOG

Status	Document Revision	Change Number	Approval Date	Description
Interim			2006-09-11	Interim Release
Baseline			2008-04-07	Baseline Release
Revision	A		2018-06-19	Significant changes were made to this NASA Technical Standard. It is recommended that it be reviewed in its entirety before implementation. Key changes were: Calibration of Eddy Current, factors to consider for CR and DR application, Level 3 or 4 penetrant sensitivity choice.

APPROVED FOR PUBLIC RELEASE – DISTRIBUTION IS UNLIMITED

NASA-STD-5009A

FOREWORD

This NASA Technical Standard is published by the National Aeronautics and Space Administration (NASA) to provide uniform engineering and technical requirements for processes, procedures, practices, and methods that have been endorsed as standard for NASA programs and projects, including requirements for selection, application, and design criteria of an item.

NASA-STD-5009A supersedes NASA-STD-5009, Nondestructive Evaluation Requirements for Fracture Critical Metallic Components, and MSFC-STD-1249, Standard NDE Guidelines and Requirements for Fracture Control Programs.

This NASA Technical Standard is approved for use by NASA Headquarters and NASA Centers and Facilities, and applicable technical requirements may be cited in contract, program, and other Agency documents. It may also apply to the Jet Propulsion Laboratory (a Federally Funded Research and Development Center (FFRDC)), other contractors, recipients of grants and cooperative agreements, and parties to other agreements only to the extent specified or referenced in applicable contracts, grants, or agreements.

This NASA Technical Standard establishes the nondestructive evaluation (NDE) requirements for any NASA system or component, flight or ground, where fracture control is a requirement. This NASA Technical Standard specifically defines requirements for nondestructive evaluation in support of NASA-STD-5019A, Fracture Control Requirements for Spaceflight Hardware.

Requests for information should be submitted via "Feedback" at https://standards.nasa.gov. Requests for changes to this NASA Technical Standard should be submitted via MSFC Form 4657, Change Request for a NASA Engineering Standard.

Original Signed by: 06/19/2018

Ralph R. Roe, Jr.
NASA Chief Engineer

Approval Date

APPROVED FOR PUBLIC RELEASE – DISTRIBUTION IS UNLIMITED

NASA-STD-5009A

TABLE OF CONTENTS

SECTION		PAGE
DOCUMENT HISTORY LOG		2
FOREWORD		3
TABLE OF CONTENTS		4
LIST OF APPENDICES		5
LIST OF FIGURES		5
LIST OF TABLES		6
1.	**SCOPE**	7
1.1	Purpose	7
1.2	Applicability	7
1.3	Tailoring	8
2.	**APPLICABLE DOCUMENTS**	8
2.1	General	8
2.2	Government Documents	8
2.3	Non-Government Documents	9
2.4	Order of Precedence	10
3.	**ACRONYMS, ABBREVIATIONS, SYMBOLS, AND DEFINITIONS**	10
3.1	Acronyms, Abbreviations, and Symbols	10
3.2	Definitions	11
4.	**NDE REQUIREMENTS**	14
4.1	NDE Procedures, Standards, and Methods	15
4.1.1	Cracks	15
4.1.2	Material Review Board (MRB)	15
4.1.3	Detailed NDE Requirements	15
4.1.4	NDE Drawing Callouts	16
4.1.5	NDE Process Documentation Control	16
4.1.6	Capability Demonstration Specimens	16
4.1.7	Supporting Data and Record Retention	17
4.1.8	Organizational Guidelines and Documentation Requirement	17
4.2	Standard NDE	17
4.2.1	Standard NDE Methods	17
4.2.2	Standard NDE Crack Sizes	17
4.2.3	Table 1 (or Table 2)—Minimum Detectable Flaw Sizes Conditional Notes	18
4.2.4	Demonstration of Standard NDE Capability	18
4.2.5	Inability to Meet Standard NDE Inspection Process Requirements	23

APPROVED FOR PUBLIC RELEASE – DISTRIBUTION IS UNLIMITED

NASA-STD-5009A

TABLE OF CONTENTS (Continued)

SECTION		PAGE
4.2.6	Standard NDE Classification Justification	24
4.2.7	Standard NDE Deviations	24
4.3	Special NDE	24
4.3.1	General	24
4.3.2	Special NDE Crack Sizes	24
4.3.3	Demonstration of Special NDE Capability	24
4.3.4	NDE Capability Demonstration Specimens	25
4.3.5	Point Estimate Method	25
4.3.6	POD Method	26
4.4	NDE Documentation	26
4.4.1	NDE Plan	26
4.4.2	NDE Summary Report	26
4.4.3	Supporting Data and Record Retention	27
4.5	Personnel Qualification and Certification	28
4.5.1	Standard NDE Qualification and Certification	28
4.5.2	Special NDE Qualification and Certification	28

LIST OF APPENDICES

APPENDIX		PAGE
A	Requirements Compliance Matrix	32
B	Example of an NDE Organization	45
C	Reference Documents	49
D	Qualification of CR and DR Methods	51

LIST OF FIGURES

FIGURE		PAGE
1	Assumed Flaw Geometries	31

APPROVED FOR PUBLIC RELEASE – DISTRIBUTION IS UNLIMITED

NASA-STD-5009A

LIST OF TABLES

TABLE		PAGE
1	Minimum Detectable Crack Sizes for Fracture Analysis Based on Standard NDE Methods (inches)	29
2	Minimum Detectable Crack Sizes for Fracture Analysis Based on Standard NDE Methods (millimeters)	30
3	System Performance Tests for CR/DR	48

NASA-STD-5009A

NONDESTRUCTIVE EVALUATION REQUIREMENTS FOR FRACTURE-CRITICAL METALLIC COMPONENTS

1. SCOPE

1.1 Purpose

The purpose of this NASA Technical Standard is to establish the nondestructive evaluation (NDE) requirements for any NASA system or component, flight or ground, where fracture control is a requirement. This NASA Technical Standard defines the primary requirements for NDE in support of NASA-STD-5019A, Fracture Control Requirements for Spaceflight Hardware. NDE applied in-process for purposes of process control and NDE of damage tolerant composites are not addressed in this document.

It is the policy of NASA to produce aerospace flight systems with a high degree of reliability and safety. This is accomplished through good design, manufacturing, test, and operational practices, including the judicious choice of materials, detailed analysis, appropriate factors of safety, rigorous testing and control of hardware, and reliable inspection. NASA fracture control requirements stipulate that all human-rated aerospace flight systems be subjected to fracture control procedures to preclude catastrophic failure. Those procedures frequently rely on NDE to ensure that the potential failure initiation of relevant crack-like flaws are not present in critical areas.

Programs that are not human-rated may choose to impose these requirements on a mission or hardware to bolster the program or to serve as a stepping-stone for human-rating.

1.2 Applicability

This NASA Technical Standard is applicable to the fracture control of metal components, e.g., aluminum, steel, titanium, and nickel alloys for any NASA system or component, flight or ground, where fracture control is a requirement.

Conditional notes on applicability are presented in section 4.2.3.

The requirements described herein apply to fracture-critical hardware developed for NASA applications by NASA Field Centers, international partners, contractors, and outside organizations. NDE processes are required to meet the requirements in this NASA Technical Standard to screen hardware reliably for the presence of crack-like flaws.

This NASA Technical Standard is approved for use by NASA Headquarters and NASA Centers and Facilities, and applicable technical requirements may be cited in contract, program, and other Agency documents. It may also apply to the Jet Propulsion Laboratory (a Federally Funded Research and Development Center (FFRDC)), other contractors, recipients of grants and

APPROVED FOR PUBLIC RELEASE – DISTRIBUTION IS UNLIMITED

cooperative agreements, and parties to other agreements only to the extent specified or referenced in applicable contracts, grants, or agreements.

Verifiable requirement statements are designated by the acronym NER (Nondestructive Evaluation Requirement), numbered and indicated by the word "shall"; this NASA Technical Standard contains 96 requirements. Explanatory or guidance text is indicated in italics beginning in section 4. To facilitate requirements selection and verification by NASA programs and projects, a Requirements Compliance Matrix is provided in Appendix A.

1.3 Tailoring

Document tailoring of the requirements in this NASA Technical Standard for application to a specific program or project as part of program or project requirements and obtain formal approval by the delegated Technical Authority in accordance with NPR 7120.5, NASA Space Flight Program and Project Management Requirements.

Technical Authority in this context may vary from program to program. In accordance with NPR 7120.10, Technical Standards for NASA Programs and Projects, section 2.2.4, "The NASA Chief Engineer, the Chief, Safety and Mission Assurance, and the Chief Health and Medical Officer serve as or may delegate Technical Authority for all technical standards within their areas of responsibility."

2. APPLICABLE DOCUMENTS

2.1 General

The documents listed in this section contain provisions that constitute requirements of this NASA Technical Standard as cited in the text.

2.1.1 The latest issuances of cited documents apply unless specific versions are designated.

2.1.2 Non-use of a specifically designated version is approved by the delegated Technical Authority.

Applicable documents may be accessed at https://standards.nasa.gov or obtained directly from the Standards Developing Body or other document distributors. When not available from these sources, information for obtaining the document is provided.

Refer to Appendix C for reference documents.

2.2 Government Documents

National Aeronautics and Space Administration

NPR 1441.1   NASA Records Management Program Requirements

APPROVED FOR PUBLIC RELEASE – DISTRIBUTION IS UNLIMITED

NPR 7120.5	NASA Program and Project Management Processes and Requirements
NPR 7120.10	Technical Standards for NASA Programs and Projects
NASA-STD-5019A	Fracture Control Requirements for Spaceflight Hardware

2.3 Non-Government Documents

Aerospace Industries Association (AIA)/National Aerospace Standards (NAS)

NAS 410	NAS Certification and Qualification of Nondestructive Test Personnel

American Society for Nondestructive Testing

Materials Evaluation, Volume 40, No. 9, 1982	"Recommended Practice for a Demonstration of Nondestructive Evaluation (NDE) Reliability on Aircraft Production Parts," Ward Rummel

American Society for Testing and Materials (ASTM)

ASTM E164	Standard Practice for Contact Ultrasonic Testing of Weldments
ASTM E1025	Standard Practice for Design, Manufacture, and Material Grouping Classification of Hole-Type Image Quality Indicators (IQI) Used for Radiology
ASTM E1316	Standard Terminology for Nondestructive Examinations
ASTM E1417/E1417M	Standard Practice for Liquid Penetrant Testing
ASTM E1444/E1444M	Standard Practice for Magnetic Particle Testing
ASTM E1742	Standard Practice for Radiographic Examination
ASTM E1817	Standard Practice for Controlling Quality of Radiological Examination by Using Representative Quality Indicators (RQIs)
ASTM E2033	Standard Practice for Radiographic Examination Using Computed Radiography (Photostimulable Luminescence Method)

APPROVED FOR PUBLIC RELEASE – DISTRIBUTION IS UNLIMITED

NASA-STD-5009A

ASTM E2375	Standard Practice for Ultrasonic Examination of Wrought Products
ASTM E2445/E2445M	Standard Practice for Performance Evaluation and Long-Term Stability of Computer Radiography Systems
ASTM E2698	Standard Practice for Radiological Examination Using Digital Detector Arrays
ASTM E2737	Standard Practice for Digital Detector Array Performance Evaluation and Long-Term Stability

SAE International

SAE AMS2647	Fluorescent Penetrant Inspection Aircraft and Engine Component Maintenance
SAE ARP4402	Eddy Current Inspection of Open Fastener Holes in Aluminum Aircraft Structure
SAE AS4787	Eddy Current Inspection of Circular Holes in Nonferrous Metallic Aircraft Engine Hardware

2.4 Order of Precedence

2.4.1 The requirements and standard practices established in this NASA Technical Standard do not supersede or waive existing requirements and standard practices found in other Agency documentation, or in applicable laws and regulations unless a specific exemption has been obtained by the Office of the NASA Chief Engineer.

2.4.2 Conflicts between this NASA Technical Standard and other requirements documents are resolved by the delegated Technical Authority.

3. ACRONYMS, ABBREVIATIONS, SYMBOLS, AND DEFINITIONS

3.1 Acronyms, Abbreviations, and Symbols

%	percent
AIA	Aerospace Industries Association
AMS	Aerospace Materials Specification
ARP	Aerospace Recommended Practice
AS	Aerospace Standard
ASTM	American Society for Testing and Materials
CR	Computed Radiology

APPROVED FOR PUBLIC RELEASE – DISTRIBUTION IS UNLIMITED

DR	Digital Radiography
FFRDC	Federally Funded Research and Development Center
HDBK	Handbook
IACS	International Annealed Copper Standard
in	inch
IQI	image quality indicator
MIL	Military
mm	Millimeter
MRB	Material Review Board
MSFC	Marshall Space Flight Center
NAS	National Aerospace Standard
NASA	National Aeronautics and Space Administration
NDE	Nondestructive Evaluation
NDI	Nondestructive Inspection
NDT	Nondestructive Testing
NIST	National Institute of Standards and Technology
NPR	NASA Procedural Requirements
NSTS	National Space Transportation System
NTIAC	Nondestructive Testing Information Analysis Center (now incorporated in the Systems Information Analysis Center)
POD	Probability of Detection
QQI	Quantitative Quality Indicator
RFCB	Responsible Fracture Control Board
RQI	Representative Quality Indicator
SI	The International System of Units (commonly known as the Système Internationale)
SSP	Space Station Program
STD	Standard

3.2 Definitions

Applicable Documents: Documents cited in the body of the standard that contain provisions or other pertinent requirements directly related and necessary to the performance of the activities specified by the standard.

Capability Demonstration Specimens: A set of specimens made from material similar to the material of the hardware to be inspected with known flaws used to estimate the capability of flaw detection, i.e., Probability of Detection (POD) or other methods of capability assessment, of an NDE system.

Certification: A written statement by an employer that an individual has met the applicable requirements of this NASA Technical Standard.

Cracks or Crack-Like Flaws: A discontinuity assumed to behave like a crack for assessment of material or structural integrity.

Defect: One or more flaws whose aggregate size, shape, orientation, location, or properties do not meet specified acceptance criteria and are rejectable.

Flaw: An imperfection or discontinuity that may be detectable by nondestructive testing and is not necessarily rejectable. Examples of flaws in metallic include cracks, deep scratches and sharp notches that behave like cracks, material inclusions, forging laps, welding incomplete fusion, penetration, and slag or porosity with a crack-like tail. For additive manufactured metallics, skipped layers, thermal or stress induced cracks, or inclusions, as examples.

Fracture Control: The rigorous application of those branches of design engineering, quality assurance, manufacturing, and operations dealing with the analysis and prevention of crack propagation leading to catastrophic failure.

Fracture-Critical Hardware, Component, or Part: Classification that assumes that cracks in the hardware, component, or part could lead to a catastrophic failure, an event that results in loss of life, serious personal injury, loss of the manned flight system, or national asset.

Hardware Developer: The organization, NASA or prime contractor, responsible for the design, development, and manufacturing of hardware that is subject to fracture control.

Initial Crack (Flaw) Size: The crack size that is assumed to exist in the part for damage tolerance analysis.

Instrument Calibration: Comparison of an instrument response with, or adjustment of an instrument response to, known references often traceable to the National Institute of Standards and Technology (NIST). This is usually performed periodically, typically at a 1-year interval. After completing calibration, a calibration sticker with calibration expiration date is affixed to the instrument.

Instrument Standardization: Adjustment of an NDE instrument response using an appropriate reference standard with known size discontinuities such as electro-discharged machined slots and flat bottom holes, to obtain or establish a known and reproducible response. This is usually done prior to an examination but can be carried out anytime there is concern about the examination or instrument response. It is also commonly known as calibration prior to initiating an NDE procedure. Instrument standardization should be carried out using a minimum of three data points demonstrating expected correlation between signal response and discontinuity size.

APPROVED FOR PUBLIC RELEASE – DISTRIBUTION IS UNLIMITED

Material Review Board (MRB): After non-conforming material has been identified, this board reviews it and determines whether or not the material should be returned, reworked, used as-is, or scrapped. Note: The MRB consists of representatives across many disciplines, including manufacturing engineering, materials engineering, Quality, etc.

Minimum Detectable Crack (Flaw) Size: The size of the smallest statistically based crack-like flaw that can be reliably detected by Standard NDE methods and that is assumed to exist in a part for the purpose of performing a damage tolerance safe-life analysis of the part, component, or assembly.

NDE Plan: A plan that describes the process for establishment, implementation, and control of NDE of aerospace flight hardware during design, manufacturing, and its operational life.

NDE Procedure: A written plan providing detailed information on "how-to" perform a hardware-specific inspection.

NDE Reference Standard: A material or object for which all relevant chemical and physical characteristics are known and measurable, used as a comparison for, or standardization of, equipment or instruments used for nondestructive testing.

90/95 Probability of Detection (POD 90/95): Refers to 90 percent probability of flaw detection with 95 percent lower confidence bound.

Nondestructive Evaluation (NDE), Nondestructive Inspection (NDI), Nondestructive Testing (NDT): The development and application of technical methods to examine materials or components in ways that do not impair future usefulness and serviceability in order to detect, locate, measure, and evaluate flaws; to assess integrity, properties, and composition; and to measure geometrical characteristics.

POD Qualification (Qualified) Flaw Size: Qualification flaw size is declared after a successful completion of POD demonstration test. It takes into account the reliably detectable flaw size, material conditions, and flaw aspect ratios. Qualification flaw sizes may be used in fracture mechanics analysis. Minimum detectable flaw size is the qualified flaw size for Standard NDE.

Qualification: The skills, training, knowledge, examinations, experience, and visual capability required for personnel to properly perform to a particular level.

Reference Document: Written, printed, or electronic matter that is useful as background information for the reader to help in understanding the subject matter but does not constitute technical requirements of the NASA Technical Standard.

Reliably Detected Flaw Size: Reliably detected flaw size is denoted as $a_{90/95}$ and implies that it can be detected with 90% POD with 95% confidence. The flaw size can be determined by successful POD demonstration test. POD Point estimation method provides a reliably detectable flaw size with minimum 90% POD with 95% confidence and is denoted as $a_{90/95}$ minimum.

Responsible Fracture Control Board (RFCB): The designated board at the NASA Center or sponsoring institution responsible for fracture control methodology that can interpret fracture control requirements. Designation may be in the form of specific duties assigned within an existing function.

Responsible NASA Center: The NASA Center where an organization or program office institutes a fracture-control program.

Responsible NDE Engineering: The NDE engineering organization of the hardware developer or the sustaining engineering organization responsible for the engineering aspect of fracture-critical NDE during manufacturing or operations and maintenance.

Special NDE: Nondestructive inspections of fracture-critical hardware that are capable of detecting cracks or crack-like flaws smaller than those assumed detectable by Standard NDE or do not conform to the requirements for Standard NDE as set forth in this document. Special NDE methods are not limited to fluorescent penetrant, radiography, ultrasonic, eddy current, and magnetic particle.

Standard NDE: NDE methods of metallic materials for which a statistically based flaw detection capability has been established. Standard NDE methods addressed by this document are limited to the fluorescent penetrant, radiographic, ultrasonic, eddy current, and magnetic particle methods employing techniques with established capabilities.

Sustaining Engineering: The organization, NASA or prime contractor, responsible for operation and maintenance of hardware that is subject to fracture control.

Technical Authority: A representative delegated by the contracting agency to address technical matters and who is responsible for the interpretation and implementation of the requirements set forth in this NASA Technical Standard.

4. NDE REQUIREMENTS

The requirements set forth in this NASA Technical Standard are the minimum NDE requirements for fracture-critical hardware. The application of Standard and Special NDE per the requirements of this NASA Technical Standard does not exempt fracture-critical hardware from routine NDE performed during manufacturing. The fracture control NDE procedures cited herein may exceed the requirements for NDE procedures that are routinely performed for purposes such as configuration control and process control. NASA-STD-5019A provides the definition of fracture-critical hardware for all spaceflight systems.

[NER 1] NASA-STD-5019A requires that all human-rated metallic fracture-critical parts shall be subjected to NDE, per this NASA Technical Standard, and defines the applications of NDE and proof testing for flaw screening.

4.1 NDE Procedures, Standards, and Methods

[NER 2] NDE procedures, standards, methods, and acceptance criteria shall be defined, validated, documented, approved, implemented, and updated during all phases of the life cycle such as manufacturing, operation, and maintenance of each fracture-critical part.

    a. [NER 3] All NDE inspections shall be conducted by certified NDE inspectors (see sections 4.5.1 and 4.5.2).

    b. [NER 4] The fracture-critical NDE inspection procedure(s) shall be clearly defined for each type of part.

    c. [NER 5] Effective and reliable NDE methods shall be selected for all part or component life cycles, including but not limited to manufacturing, maintenance, and operations.

    d. [NER 6] All identified part areas shall be inspected.

    e. [NER 7] Unless specified otherwise by the delegated Technical Authority, inspection procedures shall be designed to detect cracks and crack-like flaws in all orientations.

4.1.1 Cracks

[NER 8] All detected cracks or crack-like flaws, regardless of size, shall be reported in the NDE Summary report and dispositioned by the proper authority (see section 4.4.2.1 and section 4.4.2.2).

4.1.2 Material Review Board (MRB)

[NER 9] The acceptance of cracks of any size in a fracture-critical part shall require an MRB action and the approval of the responsible fracture control board (RFCB) and the delegated Technical Authority.

4.1.3 Detailed NDE Requirements

[NER 10] The NDE methods applied shall comply with the Standard NDE requirements of section 4.2 or the Special NDE requirements of section 4.3.

APPROVED FOR PUBLIC RELEASE – DISTRIBUTION IS UNLIMITED

4.1.4 NDE Drawing Callouts

[NER 11] NDE inspection requirements for all fracture-critical parts shall be clearly identified on all drawings.

    a. [NER 12] The drawings shall clearly identify each inspection requirement by zone when different zones require different NDE inspection requirements and acceptance criteria.

    b. [NER 13] The drawings shall be updated when NDE inspection requirements are updated.

4.1.5 NDE Process Documentation Control

    a. [NER 14] A written procedure for each fracture-critical part shall be developed that complies with the relevant specification for the NDE method selected.

    b. [NER 15] Documentation control by revision or date shall be maintained current for the following:

        (1) Personnel Qualification.
        (2) Personnel Certification.
        (3) NDE Specification.
        (4) NDE Reference Standards.
        (5) NDE Method.
        (6) NDE Part-Specific Procedures.

All NDE process changes require approval by the responsible NDE organization and delegated Technical Authority.

4.1.6 Capability Demonstration Specimens

    a. [NER 16] NDE capability demonstration specimens shall be used for determining the detection capability for all Special NDE applications and may be used to validate the capabilities of Standard NDE procedures.

    b. [NER 17] Specimens shall be representative of the material to be inspected and the critical inspection area for the applicable hardware, and of the flaw size, type, location, and orientation.

The list of parameters may vary by NDE method. Specimens may be borrowed from NASA or other Government departments when available.

    c. [NER 18] If appropriate demonstration specimens are not available, specimens shall be built or procured that meet both specimen requirements and specific engineering drawing requirements.

APPROVED FOR PUBLIC RELEASE – DISTRIBUTION IS UNLIMITED

d. [NER 19] Specimens used shall be documented as a part of the NDE procedures and personnel skill qualifications.

4.1.7 Supporting Data and Record Retention

[NER 20] All certification records, NDE reports, and associated paperwork shall be retained per NPR 1441.1, NASA Records Management Program Requirements.

4.1.8 Organizational Guidelines and Documentation Requirement

It is recommended that a document be developed that meets the intent of the responsibilities and authorities described in Appendix B.

4.2 Standard NDE

[NER 21] Standard NDE shall consist of formal nondestructive inspections of fracture-critical hardware using the NDE methods cited in Table 1, Minimum Detectable Crack Sizes for Fracture Analysis Based on Standard NDE Methods (inches), or Table 2, Minimum Detectable Crack Sizes for Fracture Analysis Based on Standard NDE Methods (millimeters).

The minimum detectable crack sizes shown in Table 1 and Table 2 can be assumed as starting points in the damage tolerance fracture analyses and are applicable only for metals. The crack geometries for the cracks in Table 1 or Table 2 are shown in Figure 1, Assumed Flaw Geometries.

4.2.1 Standard NDE Methods

[NER 22] Standard NDE methods shall be limited to eddy current, fluorescent penetrant, magnetic particle, radiography, and ultrasound.

4.2.2 Standard NDE Crack Sizes

4.2.2.1 [NER 23] Nondestructive inspections of fracture-critical hardware shall detect the initial crack sizes used in the damage tolerance fracture analyses with a capability of 90/95 (90 percent probability of detection at a 95 percent confidence level).

See section 4.2.7 for deviations from this requirement.

The minimum detectable crack sizes for the Standard NDE methods shown in Table 1 and Table 2 are based principally on an NDE capability study that was conducted on flat, fatigue-cracked 2219-T87 aluminum panels early in the Space Shuttle program, and meet the 90/95 capability requirement. Although many other similar capability studies and tests have been conducted since, in our estimation, none have universal application, neither individually nor in combination. Conducting an ideal NDE capability demonstration where all of the variables are tested is obviously unmanageable and impractical.

APPROVED FOR PUBLIC RELEASE – DISTRIBUTION IS UNLIMITED

In order to make the broadest use of NDE flaw detectability data in Table 1 or Table 2, good engineering judgment needs to be applied and should be supported by specific documented analysis of the applicability or variance. For example, a flat panel is representative of a component with a large diameter curvature. It is also reasonable to use the Table 1 or Table 2 data values for most aerospace structural alloys such as titanium or stainless steel.

4.2.2.2 *Regardless of the level of capability required of an inspection, the application of NDE generally requires that the surfaces to be inspected are clean, smooth, and accessible. Component features such as sharp radii, tight fillets, recesses, poor surface finish, poor cleanliness, non-traditional material selection, and other conditions can detrimentally influence the capability of the applied Standard NDE method.*

    a. [NER 24] If a non-ideal inspection condition is encountered, the NDE method shall be evaluated to ensure that the Standard NDE detection capability is not compromised or negatively affected (see section 4.2.3).

    b. [NER 25] The inspection condition and evaluation shall be documented in the NDE Summary Report (section 4.4.2).

4.2.3 Table 1 (or Table 2) —Minimum Detectable Flaw Sizes Conditional Notes

Since the Table 1 or Table 2 crack sizes were derived from a limited set of specimens of simple geometry, applying the crack sizes to complex geometries, other materials, material forms, material processes, and nonstandard NDE applications should be done with caution. Where the real inspection conditions deviate significantly from the concept of flat fatigue-cracked panel inspections, the transferability or similarity of the application of the Table 1 or Table 2 crack sizes to real inspection situations should be evaluated and verified by documented evidence, such as experimental data or other available test data documentation; for example, a demonstration using penetrant, where capillary action is compared using penetrant on a curved part and then a flat part. Similarity can be established by other studies, data, or by supportable rationale. Similarity considerations should meet the intent of MIL-HDBK-1823, Nondestructive Evaluation System Reliability Assessment.

If similarity cannot be established, additional tests may be required, including Standard or Special NDE demonstration tests (see section 4.3). The values listed in Table 1 and Table 2 may not apply to thick-section components; threaded parts; weldments; compressively loaded structures; double wall radiography; and other unique material, structural, or inspection applications.

4.2.4 Demonstration of Standard NDE Capability

    a. [NER 26] NDE procedure calibration or instrument standardization on simulated or real crack-like flaws shall satisfy the demonstrated detection of the minimum detectable crack size (as long as the requirements in this section for each method are also adhered to).

APPROVED FOR PUBLIC RELEASE – DISTRIBUTION IS UNLIMITED

b. [NER 27] Implementation of Standard NDE methods in accordance with the requirements in sections 4.2.4.1-4.2.4.5 shall not require a formal demonstration of crack detection capability.

4.2.4.1 Eddy Current

Standard eddy-current inspection was only applied to nonmagnetic, nonferromagnetic, and conductive metals in our early capability databases. With approval from NDE Engineering or delegated Technical Authority, inspection of some ferromagnetic materials can be considered Standard.

[NER 28] Eddy-current inspections shall be in accordance with SAE ARP4402, Eddy Current Inspection of Open Fastener Holes in Aluminum Aircraft Structure; SAE AS4787, Eddy Current Inspection of Circular Holes in Nonferrous Metallic Aircraft Engine Hardware; or NASA fracture control and NASA NDE Engineering-approved contractor internal specifications with the following additional requirements:

a. [NER 29] For part conductivities less than 10 percent International Annealed Copper Standard (IACS), the reference standard shall have a conductivity of ±2 percent IACS with a minimum reference standard conductivity of 0.8 percent IACS.

b. [NER 30] For all other part conductivities, the reference standard shall match the part conductivity within ±15 percent.

c. [NER 31] Reference standard notches used for standardization shall be no larger than the following:

(1) Rectangular Surface: 0.050 ±0.002 in (1.27 ±0.05 mm) long by 0.025 ±0.002 in (±0.05 mm) deep and ≤ 0.005 in (≤0.127 mm) wide.

(2) Triangular Corner: 0.035 ±0.002 in (±0.05 mm) long by 0.035 ±0.002 in (±0.05 mm) deep by ≤ 0.005 in (≤0.127 mm) wide.

(3) Rectangular Through Edge (part thickness ≤ 0.1 in (≤2.54 mm)): ≤ 0.1 in (≤2.54 mm) long by 0.010 ±0.001 in (±0.025 mm) deep by ≤ 0.005 in (≤0.127 mm) wide.

d. [NER 32] Noise levels on the component shall be less than 25 percent of the reference notch response.

e. [NER 33] Any indication greater than 50 percent of the reference notch response shall be reported to, and receive disposition by, the proper engineering authority.

4.2.4.1.1 [NER 34] The influence of coatings and lift-off variations on the reliability of an eddy current Standard NDE inspection process shall be evaluated for application-specific suitability and documented in the NDE Summary Report.

4.2.4.2 Fluorescent Penetrant

[NER 35] Fluorescent Penetrant inspection shall be in accordance with ASTM E1417, Standard Practice for Liquid Penetrant Testing; SAE AMS2647, Fluorescent Penetrant Inspection Aircraft and Engine Component Maintenance; or NASA fracture control and NASA NDE Engineering-approved contractor internal specifications.

4.2.4.2.1 Penetrant System

[NER 36] The penetrant system used shall be a fluorescent penetrant of Level 3 or 4 sensitivity.

(See Parker, NASA/TM-2011-215869, Jan. 2011.)

4.2.4.2.2 Mechanically Disturbed Surfaces

[NER 37] Mechanically disturbed surfaces shall be etched prior to the penetrant inspection and at an appropriate time in the manufacturing flow.

The final penetrant inspection can be performed prior to metal finishing operations such as buffing or sanding that do not by themselves produce flaws.

4.2.4.2.2.1 Etching Procedure

    a. [NER 38] An etching procedure shall be developed, approved, and controlled to prevent part damage.

    b. [NER 39] The etching procedure shall specify the minimum amount of material to be removed to ensure that smeared metal does not mask cracks.

        (1) [NER 40] Non-ferrous materials such as aluminum and titanium alloys shall be etched to remove a minimum of 0.0004 in (0.01 mm) of material.

        (2) [NER 41] Corrosion resistant steel and nickel-based alloys shall be etched to remove a minimum of 0.0002 in (0.005 mm) of material.

Care must be taken to use an etchant which will preserve the surface finish (likely a buffered etch).

    c. [NER 42] If etching is not feasible or the minimum depths are not attainable, it shall be demonstrated and documented that the required flaw size can be reliably detected following current machining processes.

4.2.4.2.2.2
[NER 43] Engineering drawing tolerances for part dimensions and finishes shall be maintained post etching.

When very close tolerances are required, critical surfaces should be machined near final dimensions, etched and inspected, then finish machined and dye penetrant inspected without further etch.

4.2.4.3 Magnetic Particle

[NER 44] Magnetic Particle inspections shall be in accordance with ASTM E1444, Standard Practice for Magnetic Particle Testing, or NASA fracture control and NASA NDE Engineering-approved contractor internal specifications.

4.2.4.3.1 [NER 45] The Magnetic Particle inspection shall be the wet, fluorescent, continuous, or multi-mag method.

4.2.4.3.2 [NER 46] A Quantitative Quality Indicator (QQI) shall be used to validate the local field intensities.

Hall probes are acceptable provided they are verified with a QQI. Pie gages are not acceptable for measuring field intensities.

4.2.4.4 Radiography (X-Ray)

Table 1, or Table 2, minimum detectable crack size for standard radiographic processes is based solely on film radiography and represents NASA's past radiographic capability demonstrations on metallic specimens. With the conversion to digital radiographic techniques and methods, it is essential to assess the capability of such systems to meet the Tables 1 and 2 requirements.

4.2.4.4.1 Film Radiography

[NER 47] Film radiographic inspections, with capabilities shown in Table 1 or Table 2, shall be in accordance with ASTM E1742, Standard Practice for Radiographic Examination, or NASA fracture control and NASA NDE Engineering-approved contractor internal specifications with the following additional requirements:

   a. [NER 48] The minimum radiographic inspection sensitivity level shall be 2-1T.

   b. [NER 49] Film density shall be 2.5 to 4.0.

   c. [NER 50] The center axis of the radiation beam shall be within ±5 degrees of the assumed crack plane orientation.

4.2.4.4.2 Digital Radiography (DR) and Computed Radiology (CR)

   a. [NER 51] Digital Radiographic (DR) inspections, with capabilities shown in Table 1 or Table 2, shall be in accordance with ASTM E2698, Standard Practice for Radiological

APPROVED FOR PUBLIC RELEASE – DISTRIBUTION IS UNLIMITED

Examination Using Digital Detector Arrays, or NASA fracture control and NASA NDE Engineering-approved contractor internal specifications.

b. [NER 52] Computed Radiographic (CR) inspections, with capabilities shown in Table 1, or Table 2, shall be in accordance with ASTM E2033, Standard Practice for Radiographic Examination Using Computed Radiography (Photostimulable Luminescence Method), or NASA fracture control and NASA NDE Engineering-approved contractor internal specifications.

c. [NER 53] For the classification of DR and CR as Standard NDE, thus meeting the capability shown in Table 1 or Table 2, each applied technique shall be qualified per a Qualification Plan approved by the responsible NDE engineering organization.

Appendix D, Table 3, System Performance Tests for CR/DR, has a list of those minimum characteristics required in the Qualification Plan for CR and DR as well as the display monitor used. In addition, the qualification plan will present the engineering approach for performing a 90/95 POD capability study on fatigue cracks.

d. [NER 54] For DR and CR, system performance parameters such as those listed in Table 3 shall be measured or checked, with the following requirements observed:

(1) [NER 55] Acceptance limits shall be established per ASTM E2445/E2445M, Standard Practice for Performance Evaluation and Long-Term Stability of Computer Radiography Systems, for CR systems and ASTM E2737, Standard Practice for Digital Detector Array Performance Evaluation and Long-Term Stability, for DR systems.

(2) [NER 56] Accompanying display monitors shall be checked per ASTM E2698.

(3) [NER 57] Frequency of checks shall be approved by the responsible NDE engineering organization.

(4) [NER 58] Minimum image quality indicator (IQI) (ASTM E1025, Standard Practice for Design, Manufacture, and Material Grouping Classification of Hole-Type Image Quality Indicators (IQI) Used for Radiology) sensitivity shall be 2-1T.

(5) [NER 59] Minimum contrast-to-noise ratio on the 1T hole shall be 2.5.

(6) [NER 60] Minimum 1T hole size-to-normalized unsharpness ratio shall be 3.

(7) [NER 61] Actual limits on contrast-to-noise ratio, 1T hole size-to-normalized unsharpness ratio and limits on x-ray angle with expected plane of the crack-like flaw shall be based upon CR and DR qualification data obtained on a representative quality indicator (RQI) per ASTM E1817, Standard Practice for

Controlling Quality of Radiological Examination by Using Representative Quality Indicators (RQIs).

    (8) [NER 62] The center axis of the radiation beam shall be within ±5 degrees of the assumed crack plane orientation.

4.2.4.5 Ultrasonics

[NER 63] Ultrasonic inspections for wrought products shall be in accordance with ASTM E2375, Standard Practice for Ultrasonic Examination of Wrought Products, ASTM E164, Standard Practice for Contact Ultrasonic Testing of Weldments, or NASA fracture control and NASA NDE Engineering-approved contractor internal specifications with the following additional requirements:

    a. [NER 64] Reference reflectors (flat bottom holes, surface notches, and side drilled holes) shall be in accordance with ASTM E2375, Class A.

    b. [NER 65] Any single discontinuity with a response greater than 50 percent of the response from a 5/64-in (1.98 mm) diameter flat bottom hole (or equivalent surface notch or side drilled hole) at the estimated discontinuity depth (or sound path) shall be reported.

    c. [NER 66] Multiple discontinuities with indications greater than the response from a 3/64-in (1.19 mm) diameter flat bottom hole (or equivalent surface notch or side drilled hole) at the estimated discontinuity depth (or sound path) shall be reported if the centers of any two such indications are separated by less than 1 in (25.4 mm).

    d. [NER 67] Any linear discontinuity (i.e., cracks or crack-like flaws such as stringers, incomplete fusion, or incomplete penetration) with a response equal to or greater than the response from a 3/64-in (1.19 mm) diameter flat bottom hole (or equivalent surface notch or side drilled hole) at the estimated discontinuity depth (or sound path) shall be reported regardless of length.

    e. [NER 68] At a minimum, shear wave inspections of welds shall be performed with the sound beam both parallel and transverse to the weld axis in two opposite directions each (4 scans total).

4.2.5 Inability to Meet Standard NDE Inspection Process Requirements

[NER 69] If the requirements of section 4.2 cannot be met, or smaller cracks or crack-like flaws than those shown in Table 1 or Table 2 have to be detected, then the inspection processes shall be considered Special NDE, and the Special NDE requirements of section 4.3 apply.

4.2.6 Standard NDE Classification Justification

4.2.6.1 [NER 70] The justification to classify an NDE procedure as a Standard NDE procedure shall be documented and then approved by NDE engineering and delegated Technical Authority.

4.2.6.2 [NER 71] The justification shall be based upon evaluation of the NDE procedure as applied to the hardware.

4.2.6.3 [NER 72] Justification shall include evaluating the similarity of the NDE procedure on the hardware with other NDE procedures on similar or identical hardware that have documented flaw detectability equal to or better than the Standard NDE minimum detectable crack size from Table 1 or Table 2.

4.2.7 Standard NDE Deviations

[NER 73] All deviations from Standard NDE shall be approved by the responsible NDE engineering organization, the RFCB, and the delegated Technical Authority.

4.3 Special NDE

4.3.1 General

Special NDE consists of nondestructive inspections that are capable of detecting crack-like flaws smaller than those detectable by Standard NDE (Table 1 or Table 2) or those that do not conform to the requirements for Standard NDE given in section 4.2, Standard NDE.

Special NDE methods are not limited to fluorescent penetrant, radiography, ultrasonic, eddy current, and magnetic particle methods.

4.3.2 Special NDE Crack Sizes

The Special NDE crack size can be any demonstrated size.

[NER 74] Special NDE inspections shall require the approval of the RFCB and the delegated Technical Authority.

4.3.3 Demonstration of Special NDE Capability

[NER 75] A 90/95 percent flaw detection capability shall be demonstrated before a Special NDE inspection can be performed for fracture-critical part screening.

The demonstration of the Special NDE inspection at a given crack size qualifies the Special NDE for implementation for the detection of cracks at the demonstrated size and larger.

a. [NER 76] The flaw detection capability of the Special NDE inspection method shall be demonstrated by testing with flawed specimens.

b. [NER 77] A sufficient number of flaws shall be included in the test demonstration to meet the 90/95 percent reliability requirement.

c. [NER 78] The tests shall be designed on the basis of either the Point Estimate Method (section 4.3.5), the Probability of Detection Method (section 4.3.6), or other approved method.

d. [NER 79] Special NDE Capability demonstration planning shall be approved by NDE engineering.

4.3.4 NDE Capability Demonstration Specimens

The most accepted method of demonstrating Special NDE capability is with fatigue-cracked specimens. The preparation and control of demonstration specimens and how to administer demonstration tests should meet the intent of MIL-HDBK-1823. In special cases, other flaws or crack types that are more representative of the application may be used for the demonstration with the approval of the RFCB and the delegated Technical Authority.

a. [NER 80] Special NDE demonstration specimen selection shall be justified and approved based on the similarity between the test hardware and the demonstration specimen.

b. [NER 81] The justification shall be documented in the NDE summary report.

4.3.5 Point Estimate Method

The Point Estimate Method approach assumes that the capability of flaw detection increases with the size of flaws in the neighborhood of the test flaw size. Since only a reliably detectable crack-like flaw size needs to be determined, a smaller number of crack-like flaws can be used to demonstrate capability.

4.3.5.1 [NER 82] The Point Estimate Method assumption shall be demonstrated or verified by documented evidence before the Point Estimate Method can be implemented.

Qualification by this method supports that the procedure and the individual operators are capable of detecting qualification size flaws with minimum 90% POD with 95% confidence. Successful point estimate demonstration involves detection of 29 flaws out of 29 flaw opportunities or minimum 45 flaws out of 46 flaw opportunities. A 90/95 minimum flaw size is given by the average flaw size in that study. (This assertion is postulated in a JSC study by A. Koshti (2015).)

4.3.5.2 [NER 83] Use of the Point Estimate Method shall be in accordance with approaches like those referenced in Materials Evaluation, Vol. 40, No. 9, pp 922-932, 1982.

APPROVED FOR PUBLIC RELEASE – DISTRIBUTION IS UNLIMITED

4.3.6 POD Method

This method requires a large range of flaw sizes that span the targeted qualification flaw size. The large number of flaws required by this method usually allows determination of a reliably detectable flaw size $a_{90/95}$. Qualification by this method demonstrates that the procedure and the individual operator are capable of reliably detecting flaws larger than or equal to the qualification flaw size. Use of the POD method should meet the intent of MIL-HDBK-1823, ASTM E3023, and ASTM E2862, as approved by the responsible NDE Engineering Organization, with the added requirements of validation of the model output and the assumptions behind each model.

[NER 84] In all POD approaches, the probability of a false call and the single-sided upper 95 percent bound shall be defined.

The false call rates for binomial analyses typically require no greater than 3.44 percent while a logit POD model requires no more than 5 percent. In the point estimate approach, this is conservatively 3 times as many blank inspection sites as there are flaw sites. In that case, it would be 87 blank inspection sites required for 29 flaws.

4.4 NDE Documentation

4.4.1 NDE Plan

[NER 85] An NDE plan shall be developed that addresses the following, as a minimum:

   a. Applicable specifications and documented standards.
   b. Calibration artifact traceability.
   c. Inspector training, qualification, and certification.
   d. Method selection, application, and process control.
   e. Acceptance criteria.
   f. Application of requirements during manufacturing, maintenance, and operations.
   g. NDE applied to fracture-critical hardware.
   h. Standard NDE selection, application, and control.
   i. Special NDE selection, equipment, application, and configuration control.

4.4.2 NDE Summary Report

[NER 86] An NDE Summary Report shall be developed and include, but not be limited to, the following:

   a. Identification of the fracture-critical part number.
   b. Critical zones inspected.
   c. NDE methods applied and procedures used.
   d. Classification and justification of Standard NDE or Special NDE inspections.

e. Acceptance criteria.
f. Inspectors' names and inspection dates.
g. Evaluation of special conditions that affect Standard NDE.

4.4.2.1 [NER 87] The report shall provide inspection results with descriptions, locations, and sizes for flaws that do not meet acceptance criteria, and other non-conformances and problems encountered during the inspection.

4.4.2.2 [NER 88] The report shall identify any cracks or crack-like flaw indications regardless of their size or disposition.

4.4.3 Supporting Data and Record Retention

a. [NER 89] The documents supporting the NDE Summary Report shall be retained in accordance with NPR 1441.1.

It is recommended that electronic and digital data be stored in common file formats that are not lost via equipment-unique data storage or limited proprietary formats.

b. [NER 90] Responsible NDE engineering shall acquire and retain all records during a change to a different contractor, including but not limited to, the following:

(1) Controlling NDE specifications and document standards.

(2) Calibration artifact traceability.

(3) Part-specific NDE procedures.

(4) Special NDE 90/95 percent capability demonstration data.

(5) Supporting data used to justify Standard NDE.

(6) Standard and Special NDE inspector qualification and certification documents.

(7) Standard and Special NDE process changes and approval documents.

(8) The hardware acceptance, inspection, and summary reports.

(9) Other supporting data, including inspector identifications, inspection dates, detailed and zoned drawings, acceptance criteria, and NDE problem reports and resolutions.

(10) NDE reports.

4.4.3.1 [NER 91] These documents shall be available for review and approval by the delegated Technical Authority, RFCB, and others.

4.5 Personnel Qualification and Certification

4.5.1 Standard NDE Qualification and Certification

[NER 92] Personnel performing Standard NDE of fracture-critical hardware shall be, at a minimum, qualified and certified Level II in accordance with NAS 410, NAS Certification and Qualification of Nondestructive Test Personnel.

4.5.2 Special NDE Qualification and Certification

a. [NER 93] Personnel performing Special NDE shall be qualified and certified for each Special NDE procedure and, as a minimum, be qualified and certified Level II in accordance with NAS 410, NAS Certification and Qualification of Nondestructive Test Personnel.

Successful demonstration of 90/95 POD on the NDE capability demonstration specimens qualifies the specific written procedure and inspector performing the inspection for detecting the demonstrated flaw size and larger.

b. [NER 94] If there is a failure to demonstrate capability, then proof of improved inspector skills shall be required prior to a retest.

c. [NER 95] Qualification for Special NDE shall be specific to the procedure and the inspector.

d. [NER 96] Special NDE inspection shall not be transferable to another procedure or inspector.

NASA-STD-5009A

Table 1—Minimum Detectable Crack Sizes for Fracture Analysis Based on Standard NDE Methods (See "Conditional Notes," section 4.2.3 for applicability.)

U. S. CUSTOMARY UNITS (inches)				
Crack Location	Part Thickness, t	Crack Type	Crack Dimension, a*	Crack Dimension, c*
Eddy Current NDE				
Open Surface	t ≤ 0.050 t > 0.050	Through PTC[1]	t 0.020 0.050	0.050 0.100 0.050
Edge or Hole	t ≤ 0.075 t > 0.075	Through Corner	t 0.075	0.100 0.075
Penetrant NDE				
Open Surface	t ≤ 0.050 0.050 < t < 0.075 t > 0.075	Through PTC	t t 0.025 0.075	0.100 0.150 - t 0.125 0.075
Edge or Hole	t ≤ 0.100 t > 0.100	Through Corner	t 0.100	0.150 0.150
Magnetic Particle NDE				
Open Surface	t ≤ 0.075 t > 0.075	Through PTC	t 0.038 0.075	0.125 0.188 0.125
Edge or Hole	t ≤ 0.075 t > 0.075	Through Corner	t 0.075	0.250 0.250
Radiographic NDE				
Open Surface	t ≤ 0.107 t > 0.107	PTC Embedded	0.7t 0.7t 2a=0.7t	0.075 0.7t 0.7t
Ultrasonic NDE Comparable to a Class A Quality Level (ASTM E2375)				
Open Surface	t ≥ 0.100	PTC Embedded**	0.030 0.065 0.017 0.039	0.150 0.065 0.087 0.039

[1] PTC - Partly through crack (Surface Crack)
* See Figure 1 for definitions of "a" and "c" for different geometries.
** Equivalent area is acceptable, ASTM E2375 Class A.

APPROVED FOR PUBLIC RELEASE – DISTRIBUTION IS UNLIMITED

Table 2—Minimum Detectable Crack Sizes for Fracture Analysis Based on Standard NDE Methods (Metric Version) (See "Conditional Notes," section 4.2.3 for applicability.)

Système International (SI) Units (millimeters)				
Crack Location	Part Thickness, t	Crack Type	Crack Dimension, a*	Crack Dimension, c*
Eddy Current NDE				
Open Surface	t ≤ 1.27 t > 1.27	Through PTC[1]	t 0.51 1.27	1.27 2.54 1.27
Edge or Hole	t ≤ 1.91 t > 1.91	Through Corner	t 1.91	2.54 1.91
Penetrant NDE				
Open Surface	t ≤ 1.27 1.27 < t < 1.91 t > 1.91	Through PTC	t t 0.64 1.91	2.54 3.81 − t 3.18 1.91
Edge or Hole	t ≤ 2.54 t > 2.54	Through Corner	t 2.54	3.81 3.81
Magnetic Particle NDE				
Open Surface	t ≤ 1.91 t > 1.91	Through PTC	t 0.97 1.91	3.18 4.78 3.18
Edge or Hole	t ≤ 1.91 t > 1.91	Through Corner	t 1.91	6.35 6.35
Radiographic NDE				
Open Surface	t ≤ 2.72 t > 2.72	PTC Embedded	0.7t 0.7t 2a=0.7t	1.91 0.7t 0.7t
Ultrasonic NDE Comparable to a Class A Quality Level (ASTM E2375)				
Open Surface	t ≥ 2.54	PTC Embedded**	0.76 1.65 0.43 0.99	3.81 1.65 2.21 0.99

[1] PTC – Partly through crack (Surface Crack)
* See Figure 1 for definitions of "a" and "c" for different geometries.
** Equivalent area is acceptable, ASTM E2375 Class A.

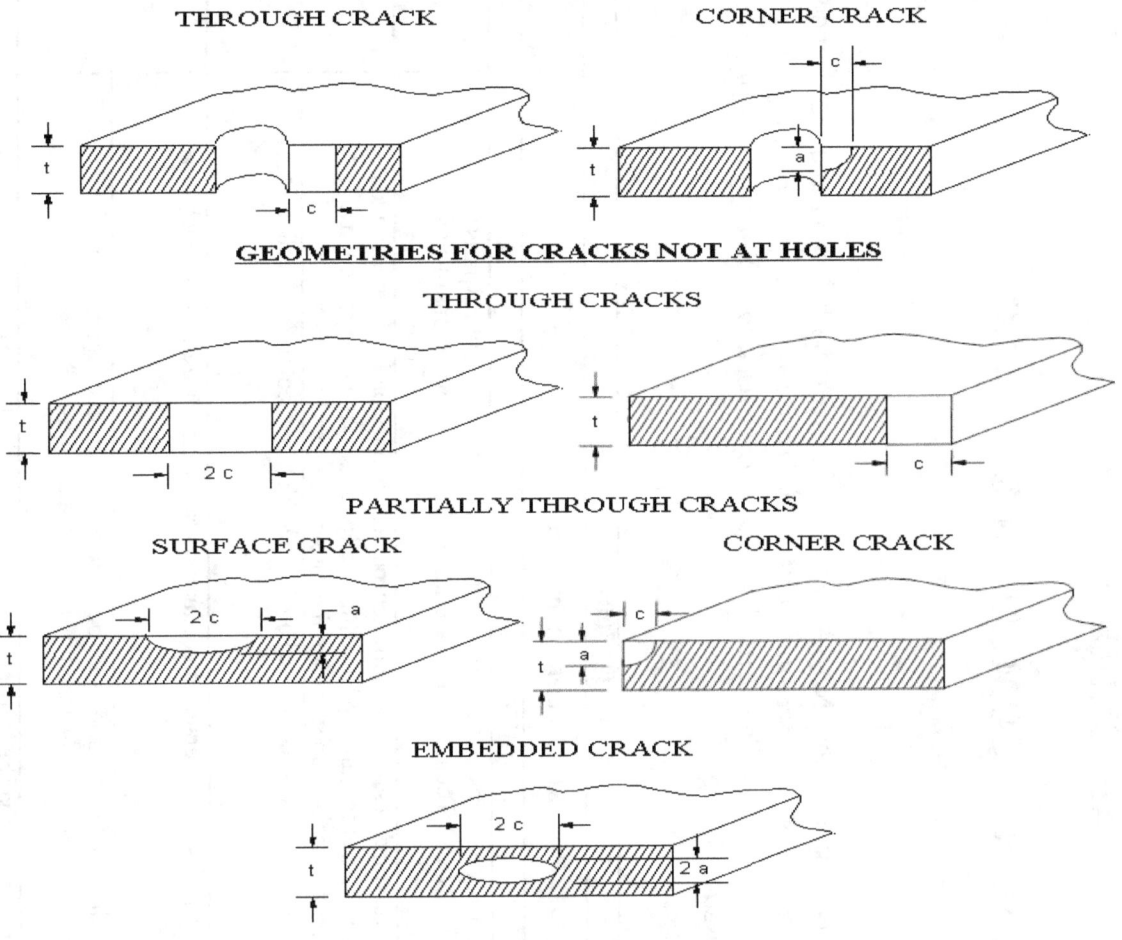

Figure 1—Assumed Flaw Geometries

NASA-STD-5009A

APPENDIX A

REQUIREMENTS COMPLIANCE MATRIX

A.1 Purpose

This Appendix provides a listing of requirements contained in this NASA Technical Standard for selection and verification of requirements by programs and projects. (*Note*: Enter "Yes" if the requirement is applicable to the program or project or "No" if the requirement is not applicable to the program or project. The "Comments" column may be used to provide specific instructions on how to apply the requirement or to specify proposed tailoring.)

Section	Description	NASA-STD-5009A Requirement in this Standard	Applicable (Yes or No)	Comments
		4. NDE Requirements		
4.	NDE Requirements	[NER 1] NASA-STD-5019A requires that all human-rated metallic fracture-critical parts shall be subjected to NDE, per this NASA Technical Standard, and defines the applications of NDE and proof testing for flaw screening.		
4.1	NDE Procedures, Standards, and Methods	[NER 2] NDE procedures, standards, methods, and acceptance criteria shall be defined, validated, documented, approved, implemented, and updated during all phases of the life cycle such as manufacturing, operation, and maintenance of each fracture-critical part.		
4.1a	NDE Procedures, Standards, and Methods	[NER 3] All NDE inspections shall be conducted by certified NDE inspectors (see sections 4.5.1 and 4.5.2).		
4.1b	NDE Procedures, Standards, and Methods	[NER 4] The fracture-critical NDE inspection procedure(s) shall be clearly defined for each type of part.		
4.1c	NDE Procedures, Standards, and Methods	[NER 5] Effective and reliable NDE methods shall be selected for all part or component life cycles, including but not limited to manufacturing, maintenance, and operations.		
4.1d	NDE Procedures, Standards, and Methods	[NER 6] All identified part areas shall be inspected.		

APPROVED FOR PUBLIC RELEASE – DISTRIBUTION IS UNLIMITED

NASA-STD-5009A

Section	Description	Requirement in this Standard	Applicable (Yes or No)	Comments
4.1e	NDE Procedures, Standards, and Methods	[NER 7] Unless specified otherwise by the delegated Technical Authority, inspection procedures shall be designed to detect cracks and crack-like flaws in all orientations.		
4.1.1	Cracks	[NER 8] All detected cracks or crack-like flaws, regardless of size, shall be reported in the NDE Summary report and dispositioned by the proper authority (see section 4.4.2.1 and section 4.4.2.2).		
4.1.2	Material Review Board (MRB)	[NER 9] The acceptance of cracks of any size in a fracture-critical part shall require an MRB action and the approval of the responsible fracture control board (RFCB) and the delegated Technical Authority.		
4.1.3	Detailed NDE Requirements	[NER 10] The NDE methods applied shall comply with the Standard NDE requirements of section 4.2 or the Special NDE requirements of section 4.3.		
4.1.4	NDE Drawing Callouts	[NER 11] NDE inspection requirements for all fracture-critical parts shall be clearly identified on all drawings.		
4.1.4a	NDE Drawing Callouts	[NER 12] The drawings shall clearly identify each inspection requirement by zone when different zones require different NDE inspection requirements and acceptance criteria.		
4.1.4b	NDE Drawing Callouts	[NER 13] The drawings shall be updated when NDE inspection requirements are updated.		
4.1.5	NDE Process Documentation Control	[NER 14] A written procedure for each fracture-critical part shall be developed that complies with the relevant specification for the NDE method selected.		
4.1.5a	NDE Process Documentation Control	[NER 15] Documentation control by revision or dates shall be maintained current for the following: (1) Personnel Qualification. (2) Personnel Certification. (3) NDE Specification. (4) NDE Reference Standards. (5) NDE Method (6) NDE Part-Specific Procedures.		
4.1.6a	Capability Demonstration Specimens	[NER 16] NDE capability demonstration specimens shall be used for determining the detection capability for all Special NDE applications and may be used to validate the capabilities of Standard NDE procedures.		
4.1.6b	Capability Demonstration Specimens	[NER 17] Specimens shall be representative of the material to be inspected and the critical inspection area for the applicable hardware, and of the flaw size, type, location, and orientation.		
4.1.6c	Capability Demonstration Specimens	[NER 18] If appropriate demonstration specimens are not available, specimens shall be built or procured that meet both specimen requirements and specific engineering drawing requirements.		

APPROVED FOR PUBLIC RELEASE – DISTRIBUTION IS UNLIMITED

NASA-STD-5009A

Section	Description	Requirement in this Standard	Applicable (Yes or No)	Comments
4.1.6d	Capability Demonstration Specimens	[NER 19] Specimens used shall be documented as a part of the NDE procedures and personnel skill qualifications.		
4.1.7	Supporting Data an Record Retention	[NER 20] All certification records, NDE reports, and associated paperwork shall be retained per NPR 1441.1, NASA Records Management Program Requirements.		
4.2	Standard NDE	[NER 21] Standard NDE shall consist of formal nondestructive inspections of fracture-critical hardware using the NDE methods cited in Table 1, Minimum Detectable Crack Sizes for Fracture Analysis Based on Standard NDE Methods (inches), or Table 2, Minimum Detectable Crack Sizes for Fracture Analysis Based on Standard NDE Methods (millimeters).		

Table 1—Minimum Detectable Crack Sizes for Fracture Analysis Based on Standard NDE Methods (See "Conditional Notes," section 4.2.3 for applicability.)

U.S. CUSTOMARY UNITS (inches)

Crack Location	Part Thickness, t	Crack Type	Crack Dimension, a*	Crack Dimension, c*
Eddy Current NDE				
Open Surface	t ≤ 0.050	Through PTC[1]	t	0.050
	0.050 t > 0.050		0.020	0.100
			0.050	0.050
Edge or Hole	t ≤ 0.075	Through Corner	t	0.100
			0.075	0.075
Penetrant NDE				
Open Surface	t ≤ 0.050	Through PTC	t	0.100
	0.050<t <0.075		t	0.150 - t
	t > 0.075		0.025	0.125
			0.075	0.075
Edge or Hole	t ≤ 0.100	Through Corner	t	0.150
			0.100	0.150

NASA-STD-5009A

Section	Description	Applicable (Yes or No)	Comments
	NASA-STD-5009A **Requirement in this Standard** **Table 2—Minimum Detectable Crack Sizes for Fracture Analysis Based on Standard NDE Methods (Metric Version) (See "Conditional Notes," section 4.2.3 for applicability.)** **Système International (SI) Units (millimeters)** 		

Crack Location	Part Thickness, t	Crack Type	Crack Dimension, a*	Crack Dimension, c*
Magnetic Particle NDE				
Open Surface	t ≤ 0.075 t > 0.075	Through PTC	t 0.038 0.075	0.125 0.188 0.125
Edge or Hole	t ≤ 0.075 t	Through Corner	t 0.075	0.250 0.250
Radiographic NDE				
Open Surface	t ≤ 0.107 t > 0.107	PTC Embedded	0.7t 0.7t 2a=0.7t	0.075 0.7t 0.7t
Ultrasonic NDE Comparable to a Class A Quality Level (ASTM				
Open Surface	t ≥ 0.100	PTC Embedded**	0.030 0.065 0.017 0.039	0.150 0.065 0.087 0.039

¹ PTC - Partly through crack (Surface Crack)
* See Figure 1 for definitions of "a" and "c" for different geometries.
** Equivalent area is acceptable, ASTM E2375 Class A.

APPROVED FOR PUBLIC RELEASE – DISTRIBUTION IS UNLIMITED

NASA-STD-5009A

Section	Description	Requirement in this Standard			Applicable (Yes or No)	Comments
		Eddy Current NDE				
		Open Surface	t ≤ 1.27 t > 1.27	Through PTC[1]	t 0.51 1.27	1.27 2.54 1.27
		Edge or Hole	t ≤ 1.91 t > 1.91	Through Corner	t 1.91	2.54 1.91
		Penetrant NDE				
		Open Surface	t ≤ 1.27 1.27 < t < 1.91 t > 1.91	Through PTC	t t 0.64 1.91	2.54 3.81 – t 3.18 1.91
		Edge or Hole	t ≤ 2.54 t > 2.54	Through Corner	t 2.54	3.81 3.81
		Magnetic Particle NDE				
		Open Surface	t ≤ 1.91 t > 1.91	Through PTC	t 0.97 1.91	3.18 4.78 3.18
		Edge or Hole	t ≤ 1.91 t > 1.91	Through Corner	t 1.91	6.35 6.35
		Radiographic NDE				
		Open Surface	t ≤ 2.72 t > 2.72	PTC Embedded	0.7t 0.7t 2a=0.7t	1.91 0.7t 0.7t
		Ultrasonic NDE				
		Comparable to a Class A Quality Level (ASTM E2375)				

NASA-STD-5009A

Section	Description	Requirement in this Standard			Applicable (Yes or No)	Comments
		Open Surface	PTC¹	Embedded**		
		t ≥ 2.54	0.76 / 1.65 / 0.43 / 0.99	3.81 / 1.65 / 2.21 / 0.99		
		¹PTC – Partly through crack (Surface Crack) * See Figure 1 for definitions of "a" and "c" for different geometries. ** Equivalent area is acceptable, ASTM E2375 Class A.				
4.2.1	Standard NDE Methods	[NER 22] Standard NDE methods shall be limited to eddy current, fluorescent penetrant, magnetic particle, radiography, and ultrasound.				
4.2.2.1	Standard NDE Crack Sizes	[NER 23] Nondestructive inspections of fracture-critical hardware shall detect the initial crack sizes used in the damage tolerance fracture analyses with a capability of 90/95 (90 percent probability of detection at a 95 percent confidence level).				
4.2.2.2a	Standard NDE Crack Sizes	[NER 24] If a non-ideal inspection condition is encountered, the NDE method shall be evaluated to ensure that the Standard NDE detection capability is not compromised or negatively affected (see section 4.2.3).				
4.2.2.2b	Standard NDE Crack Sizes	[NER 25] The inspection condition and evaluation shall be documented in the NDE Summary Report (section 4.4.2).				
4.2.4a	Demonstration of Standard NDE Capability	[NER 26] NDE procedure calibration or instrument standardization on simulated or real crack-like flaws shall satisfy the demonstrated detection of the minimum detectable crack size (as long as the requirements in this section for each method are also adhered to).				
4.2.4b	Demonstration of Standard NDE Capability	[NER 27] Implementation of Standard NDE methods in accordance with the requirements in sections 4.2.4.1-4.2.4.5 shall not require a formal demonstration of crack detection capability.				
4.2.4.1	Eddy Current	[NER 28] Eddy-current inspections shall be in accordance with SAE ARP4402, Eddy Current Inspection of Open Fastener Holes in Aluminum Aircraft Structure; SAE AS4787, Eddy Current Inspection of Circular Holes in Nonferrous Metallic Aircraft Engine Hardware; or NASA fracture control and NASA NDE Engineering-approved contractor internal specifications with the following additional requirements:				
4.2.4.1a	Eddy Current	[NER 29] For part conductivities less than 10 percent International Annealed Copper Standard (IACS), the reference standard shall have a conductivity of ±2 percent IACS with a minimum reference standard conductivity of 0.8 percent IACS.				
4.2.4.1b	Eddy Current	[NER 30] For all other part conductivities, the reference standard shall match the part conductivity within ±15 percent.				
4.2.4.1c	Eddy Current	[NER 31] Reference standard notches used for standardization shall be no larger than the following:				

NASA-STD-5009A

Section	Description	Requirement in this Standard	Applicable (Yes or No)	Comments
		(1) Rectangular Surface: 0.050 ±0.002 in (1.27 ±0.05 mm) long by 0.025 ±0.002 in (±0.05 mm) deep and ≤0.005 in (≤0.127 mm) wide.		
		(2) Triangular Corner: 0.035 ±0.002 in (±0.05 mm) long by 0.035 ±0.002 in (±0.05 mm) deep by ≤0.005 in (≤0.127 mm) wide.		
		(3) Rectangular Through Edge (part thickness ≤0.1 in (≤2.54 mm)): ≤0.1 in (≤2.54 mm) long by 0.010 ±0.001 in (±0.025 mm) deep by ≤0.005 in (≤0.127 mm) wide.		
4.2.4.1d	Eddy Current	[NER 32] Noise levels on the component shall be less than 25 percent of the reference notch response.		
4.2.4.1e	Eddy Current	[NER 33] Any indication greater than 50 percent of the reference notch response shall be reported to, and receive disposition by, the proper engineering authority.		
4.2.4.1.1	Eddy Current	[NER 34] The influence of coatings and lift-off variations on the reliability of an eddy current Standard NDE inspection process shall be evaluated for application-specific suitability and documented in the NDE Summary Report.		
4.2.4.2	Fluorescent Penetrant	[NER 35] Fluorescent Penetrant inspection shall be in accordance with ASTM E1417, Standard Practice for Liquid Penetrant Testing; SAE AMS2647, Fluorescent Penetrant Inspection Aircraft and Engine Component Maintenance; or NASA fracture control and NASA NDE Engineering- approved contractor internal specifications.		
4.2.4.2.1	Penetrant System	[NER 36] The penetrant system used shall be a fluorescent penetrant of Level 3 or 4 sensitivity.		
4.2.4.2.2	Mechanically Disturbed Surfaces	[NER 37] Mechanically disturbed surfaces shall be etched prior to the penetrant inspection and at an appropriate time in the manufacturing flow.		
4.2.4.2.2.1a	Etching Procedure	[NER 38] An etching procedure shall be developed, approved, and controlled to prevent part damage.		
4.2.4.2.2.1b	Etching Procedure	[NER 39] The etching procedure shall specify the minimum amount of material to be removed to ensure that smeared metal does not mask cracks.		
4.2.4.2.2.1b(1)	Etching Procedure	[NER 40] Non-ferrous materials such as aluminum and titanium alloys shall be etched to remove a minimum of 0.0004 in (0.01 mm) of material.		
4.2.4.2.2.1b(2)	Etching Procedure	[NER 41] Corrosion resistant steel and nickel-based alloys shall be etched to remove a minimum of 0.0002 in (0.005 mm) of material.		
4.2.4.2.2.1c	Etching Procedure	[NER 42] If etching is not feasible or the minimum depths are not attainable, it shall be demonstrated and documented that the required flaw size can be reliably detected following current machining processes.		
4.2.4.2.2.2	Etching Procedure	[NER 43] Engineering drawing tolerances for part dimensions and finishes shall be maintained post etching.		

NASA-STD-5009A

Section	Description	Requirement in this Standard	Applicable (Yes or No)	Comments
4.2.4.3	Magnetic Particle	[NER 44] Magnetic Particle inspections shall be in accordance with ASTM E1444, Standard Practice for Magnetic Particle Testing, or NASA fracture control and NASA NDE Engineering -approved contractor internal specifications.		
4.2.4.3.1	Magnetic Particle	[NER 45] The Magnetic Particle inspection shall be the wet, fluorescent, continuous, or multi-mag method.		
4.2.4.3.2	Magnetic Particle	[NER 46] A Quantitative Quality Indicator (QQI) shall be used to validate the local field intensities.		
4.2.4.4.1	Film Radiography	[NER 47] Film radiographic inspections, with capabilities shown in Table 1 or Table 2, shall be in accordance with ASTM E1742, Standard Practice for Radiographic Examination, or NASA fracture control and NASA NDE Engineering -approved contractor internal specifications with the following additional requirements:		
4.2.4.4.1a	Film Radiography	[NER 48] The minimum radiographic inspection sensitivity level shall be 2-1T.		
4.2.4.4.1b	Film Radiography	[NER 49] Film density shall be 2.5 to 4.0.		
4.2.4.4.1c	Film Radiography	[NER 50] The center axis of the radiation beam shall be within ±5 degrees of the assumed crack plane orientation.		
4.2.4.4.2a	Digital Radiography (DR) and Computed Radiology (CR)	[NER 51] Digital Radiographic (DR) inspections, with capabilities shown in Table 1 or Table 2, shall be in accordance with ASTM E2698, Standard Practice for Radiological Examination Using Digital Detector Arrays, or NASA fracture control and NASA NDE Engineering -approved contractor internal specifications.		
4.2.4.4.2b	Digital Radiography (DR) and Computed Radiology (CR)	[NER 52] Computed Radiographic (CR) inspections, with capabilities shown in Table 1, or Table 2, shall be in accordance with ASTM E2033, Standard Practice for Radiographic Examination Using Computed Radiography (Photostimulable Luminescence Method), or NASA fracture control and NASA NDE Engineering-approved contractor internal specifications.		
4.2.4.4.2c	Digital Radiography (DR) and Computed Radiology (CR)	[NER 53] For the classification of DR and CR as Standard NDE, each applied technique shall be qualified per a Qualification Plan approved by the responsible NDE engineering organization, Table 1 or Table 2, thus meeting the capability shown in		
4.2.4.4.2d	Digital Radiography (DR) and Computed Radiology (CR)	[NER 54] For DR and CR, system performance parameters such as those listed in Table 3 shall be measured or checked, with the following requirements observed:		
4.2.4.4.2d(1)	Digital Radiography (DR) and Computed Radiology (CR)	[NER 55] Acceptance limits shall be established per ASTM E2445/E2445M, Standard Practice for Performance Evaluation and Long-Term Stability of Computer Radiography Systems, for CR systems and ASTM E2737, Standard Practice for Digital Detector Array Performance Evaluation and Long-Term Stability, for DR systems.		

APPROVED FOR PUBLIC RELEASE – DISTRIBUTION IS UNLIMITED

NASA-STD-5009A

Section	Description	Requirement in this Standard	Applicable (Yes or No)	Comments
4.2.4.4.2d(2)	Digital Radiography (DR) and Computed Radiology (CR)	[NER 56] Accompanying display monitors shall be checked per ASTM E2698.		
4.2.4.4.2d(3)	Digital Radiography (DR) and Computed Radiology (CR)	[NER 57] Frequency of checks shall be approved by the responsible NDE engineering organization.		
4.2.4.4.2d(4)	Digital Radiography (DR) and Computed Radiology (CR)	[NER 58] Minimum image quality indicator (IQI) (ASTM E1025, Standard Practice for Design, Manufacture, and Material Grouping Classification of Hole-Type Image Quality Indicators (IQI) Used for Radiology) sensitivity shall be 2-1T.		
4.2.4.4.2d(5)	Digital Radiography (DR) and Computed Radiology (CR)	[NER 59] Minimum contrast-to-noise ratio on the 1T hole shall be 2.5.		
4.2.4.4.2d(6)	Digital Radiography (DR) and Computed Radiology (CR)	[NER 60] Minimum 1T hole size-to-normalized unsharpness ratio shall be 3.		
4.2.4.4.2d(7)	Digital Radiography (DR) and Computed Radiology (CR)	[NER 61] Actual limits on contrast-to-noise ratio, 1T hole size-to-normalized unsharpness ratio and limits on x-ray angle with expected plane of the crack-like flaws shall be based upon CR and DR qualification data obtained on a representative quality indicator (RQI) per ASTM E1817, Standard Practice for Controlling Quality of Radiological Examination by Using Representative Quality Indicators (RQIs).		
4.2.4.4.2d(8)	Digital Radiography (DR) and Computed Radiology (CR)	[NER 62] The center axis of the radiation beam shall be within ±5 degrees of the assumed crack plane orientation.		
4.2.4.5	Ultrasonics	[NER 63] Ultrasonic inspections for wrought products shall be in accordance with ASTM E2375, Standard Practice for Ultrasonic Examination of Wrought Products, ASTM E164, Standard Practice for Contact Ultrasonic Testing of Weldments, or NASA fracture control and NASA NDE Engineering-approved contractor internal specifications with the following additional requirements:		
4.2.4.5a	Ultrasonics	[NER 64] Reference reflectors (flat bottomholes, surface notches, and side drilled holes) shall be in accordance with ASTM E2375, Class A.		
4.2.4.5b	Ultrasonics	[NER 65] Any single discontinuity with a response greater than 50 percent of the response from a 5/64-in (1.98 mm) diameter flat bottomhole (or equivalent surface notch or side drilled hole) at the estimated discontinuity depth (or sound path) shall be reported.		
4.2.4.5c	Ultrasonics	[NER 66] Multiple discontinuities with indications greater than the response from a 3/64-in (1.19 mm) diameter flat bottomhole (or equivalent surface notch or side drilled hole) at the estimated		

NASA-STD-5009A

Section	Description	Requirement in this Standard	Applicable (Yes or No)	Comments
		NASA-STD-5009A		
4.2.4.5d	Ultrasonics	discontinuity depth (or sound path) shall be reported if the centers of any two such indications are separated by less than 1 in (25.4 mm).		
	Ultrasonics	[NER 67] Any linear discontinuity (i.e., cracks or crack-like flaws such as stringers, incomplete fusion, or incomplete penetration) with a response equal to or greater than the response from a 3/64-in (1.19 mm) diameter flat bottom hole (or equivalent surface notch or side drilled hole) at the estimated discontinuity depth (or sound path) shall be reported regardless of length.		
4.2.4.5e	Ultrasonics	[NER 68] At a minimum, shear wave inspections of welds shall be performed with the sound beam both parallel and transverse to the weld axis in two opposite directions each (4 scans total).		
4.2.5	Inability to Meet Standard NDE Inspection Process Requirements	[NER 69] If the requirements of section 4.2 cannot be met, or smaller cracks or crack-like flaws than those shown in Table 1 or Table 2 have to be detected, then the inspection processes shall be considered Special NDE, and the Special NDE requirements of section 4.3 apply.		
4.2.6.1	Standard NDE Classification Justification	[NER 70] The justification to classify an NDE procedure as a Standard NDE procedure shall be documented and then approved by NDE engineering and the delegated Technical Authority.		
4.2.6.2	Standard NDE Classification Justification	[NER 71] The justification shall be based upon evaluation of the NDE procedure as applied to the hardware.		
4.2.6.3	Standard NDE Classification Justification	[NER 72] Justification shall include evaluating the similarity of the NDE procedure on the hardware with other NDE procedures on similar or identical hardware that have documented flaw detectability equal to or better than the Standard NDE minimum detectable crack size from Table 1 or Table 2.		
4.2.7	Standard NDE Deviations	[NER 73] All deviations from Standard NDE shall be approved by the responsible NDE engineering organization, the RFCB, and the delegated Technical Authority.		
4.3.2	Special NDE Crack Sizes	[NER 74] Special NDE inspections shall require the approval of the RFCB and the delegated Technical Authority.		
4.3.3	Demonstration of Special NDE Capability	[NER 75] A 90/95 percent flaw detection capability shall be demonstrated before a Special NDE inspection can be performed for fracture-critical part screening.		
4.3.3a	Demonstration of Special NDE Capability	[NER 76] The flaw detection capability of the Special NDE inspection method shall be demonstrated by testing with flawed specimens.		
4.3.3b	Demonstration of Special NDE Capability	[NER 77] A sufficient number of flaws shall be included in the test demonstration to meet the 90/95 percent reliability requirement.		

APPROVED FOR PUBLIC RELEASE – DISTRIBUTION IS UNLIMITED

NASA-STD-5009A

Section	Description	Requirement in this Standard	Applicable (Yes or No)	Comments
4.3.3c	Demonstration of Special NDE Capability	[NER 78] The tests shall be designed on the basis of either the Point Estimate Method (section 4.3.5), the Probability of Detection Method (section 4.3.6), or other approved method.		
4.3.3d	Demonstration of Special NDE Capability	[NER 79] Special NDE Capability demonstration planning shall be approved by NDE engineering.		
4.3.4a	NDE Capability Demonstration Specimens	[NER 80] Special NDE demonstration specimen selection shall be justified and approved based on the similarity between the test hardware and the demonstration specimen.		
4.3.4b	NDE Capability Demonstration Specimens	[NER 81] The justification shall be documented in the NDE summary report.		
4.3.5.1	Point Estimate Method	[NER 82] The Point Estimate Method assumption shall be demonstrated or verified by documented evidence before the Point Estimate Method can be implemented.		
4.3.5.2	Point Estimate Method	[NER 83] Use of the Point Estimate Method shall be in accordance with approaches like those referenced in Materials Evaluation, Vol. 40, No. 9, pp 922-932, 1982.		
4.3.6	POD Method	[NER 84] In all POD approaches, the probability of a false call and the single-sided upper 95 percent bound shall be defined.		
4.4.1	NDE Plan	[NER 85] An NDE plan shall be developed that addresses the following, as a minimum: a. Applicable specifications and standards. b. Calibration artifact traceability. c. Inspector training, qualification, and certification. d. Method selection, application, and process control. e. Acceptance criteria. f. Application of requirements during manufacturing, maintenance, and operations. g. NDE applied to fracture-critical hardware. h. Standard NDE selection, application, and control. i. Special NDE selection, equipment, application, and configuration control.		
4.4.2	NDE Summary Report	[NER 86] An NDE Summary Report shall be developed and include, but not be limited to, the following: a. Identification of the fracture-critical part number. b. Critical zones inspected. c. NDE methods applied and procedures used.		

APPROVED FOR PUBLIC RELEASE – DISTRIBUTION IS UNLIMITED

NASA-STD-5009A

Section	Description	Requirement in this Standard	Applicable (Yes or No)	Comments
		d. Classification and justification of Standard NDE or Special NDE inspections. e. Acceptance criteria. f. Inspectors' names and inspection dates. g. Evaluation of special conditions that affect Standard NDE.		
4.4.2.1	NDE Summary Report	[NER 87] The report shall provide inspection results with descriptions, locations, and sizes for flaws that do not meet acceptance criteria, and other non-conformances and problems encountered during the inspection.		
4.4.2.2	NDE Summary Report	[NER 88] The report shall identify any cracks or crack-like flaw indications regardless of their size or disposition.		
4.4.3a	Supporting Data an Record Retention	[NER 89] The documents supporting the NDE Summary Report shall be retained in accordance with NPR 1441.1.		
4.4.3b	Supporting Data an Record Retention	[NER 90] Responsible NDE engineering shall acquire and retain all records during a change to a different contractor, including but not limited to, the following: (1) Controlling NDE specifications and document standards. (2) Calibration artifact traceability. (3) Part-specific NDE procedures. (4) Special NDE 90/95 percent capability demonstration data. (5) Supporting data used to justify Standard NDE. (6) Standard and Special NDE inspector qualification and certification documents. (7) Standard and Special NDE process changes and approval documents. (8) The hardware acceptance, inspection, and summary reports. (9) Other supporting data, including inspector identifications, inspection dates, detailed and zoned drawings, acceptance criteria, and NDE problem reports and resolutions. (10) NDE reports.		
4.4.3.1	Supporting Data an Record Retention	[NER 91] These documents shall be available for review and approval by the delegated Technical Authority, RFCB, and others.		
4.5.1	Standard NDE Qualification and Certification	[NER 92] Personnel performing Standard NDE of fracture-critical hardware shall be, at a minimum, qualified and certified Level II in accordance with NAS 410, NAS Certification and Qualification of Nondestructive Test Personnel.		
4.5.2a	Special NDE Qualification and Certification	[NER 93] Personnel performing Special NDE shall be qualified and certified for each Special NDE procedure and, as a minimum, be qualified and certified Level II in accordance with NAS 410, NAS Certification and Qualification of Nondestructive Test Personnel.		
4.5.2b	Special NDE Qualification and Certification	[NER 94] If there is a failure to demonstrate capability, then proof of improved inspector skills shall be required prior to a retest.		

APPROVED FOR PUBLIC RELEASE – DISTRIBUTION IS UNLIMITED

NASA-STD-5009A

Section	Description	NASA-STD-5009A Requirement in this Standard	Applicable (Yes or No)	Comments
4.5.2c	Special NDE Qualification and Certification	[NER 95] Qualification for Special NDE shall be specific to the procedure and the inspector.		
4.5.2d	Special NDE Qualification and Certification	[NER 96] Special NDE inspection shall not be transferable to another procedure or inspector.		

APPROVED FOR PUBLIC RELEASE – DISTRIBUTION IS UNLIMITED

NASA-STD-5009A

APPENDIX B

EXAMPLE OF AN NDE ORGANIZATION ("SHALLS" ARE FOR EXAMPLE ONLY)

This Appendix provides, in what follows, an example of an NDE organization with clearly defined roles, responsibilities, and implementation. This is not a requirement and may be used as an organizational model.

NOTE: The "shalls" used below are for example purposes only and are not requirements related to this NASA Technical Standard.

B.1 Implementation and Responsibilities

B.1.1 Oversight Responsibility

The responsible NASA Center shall provide an NDE oversight function and approval of the hardware developer's NDE plan and program for the inspection of fracture-critical hardware.

a. The NASA oversight function shall establish periodic reviews of the hardware developer's NDE program.

b. The NDE oversight function shall be responsible for establishing the personnel certification system for Special NDE procedures when NASA conducts the certification tests.

c. NDE standards and procedures for fracture-critical hardware shall be approved by the responsible NASA Center.

d. In general, all plans, data, documentation, reference standards, and reliability demonstration specimens generated under contract from NASA to its contractors, subcontractors, and suppliers in fulfillment of these requirements during hardware development, manufacturing, operations, and maintenance shall be subject to examination, evaluation, and inspection by and delivery to the NDE oversight function or designated representatives of the responsible NASA Center.

B.1.2 Responsibility of Hardware-Specific NDE Requirements

a. The hardware developer, in concurrence with the RFCB and Technical Authority, shall establish and provide hardware-specific NDE requirements to its in-house NDE inspection organizations, suppliers, subcontractors, and vendors to accomplish the NDE during manufacturing.

APPROVED FOR PUBLIC RELEASE – DISTRIBUTION IS UNLIMITED

b. The hardware developer, in concurrence with the RFCB and Technical Authority, shall provide the hardware-specific NDE requirements for operations and maintenance to the responsible NASA Center or its designated sustaining engineering organization.

c. The sustaining engineering organization, in concurrence with the RFCB and Technical Authority, shall be responsible for maintaining, changing, and establishing new hardware-specific fracture control requirements.

d. The hardware developer or sustaining engineering organization shall perform a drawing review process that identifies fracture-critical parts, identifies all areas of the parts requiring NDE, and identifies the appropriate type of NDE during manufacturing, maintenance, and operations.

e. The review process shall include NDE engineering.

f. NDE engineering shall ensure that the identified areas are inspectable and that efficient and reliable NDE methods are selected.

B.1.3 Responsibility of NDE Standards, Procedures, and Reference Standards During Hardware Development and Manufacturing

a. The hardware developer's responsible NDE engineering shall be responsible for establishing and approving NDE method standards, NDE procedures, and NDE reference standards and for ensuring that all NDE processes are implemented through approved written NDE procedures.

b. The NDE procedures shall be performed by the hardware developer's NDE inspection organization or by an approved external NDE organization, provided that the NDE procedures approved by responsible NDE engineering are used by personnel certified for fracture-critical NDE.

c. NDE engineering shall be responsible for administering Special NDE certification tests and approving certification of the NDE personnel.

B.1.4 Responsibility of NDE Standards, Procedures, and Reference Standards for Operations and Maintenance NDE

a. During hardware development, the hardware developer shall be responsible for establishing, approving, and providing operations and maintenance NDE requirements, method standards, NDE procedures, and NDE reference standards to the sustaining engineering organization designated by the responsible NASA Center.

b. During operations and maintenance, the sustaining engineering organization, in concurrence with the RFCB and Technical Authority, shall be responsible for maintaining or changing existing NDE requirements, standards, and procedures as well as establishing and maintaining new hardware-specific NDE requirements, standards, and procedures. *NDE procedures may be performed by an in-house NDE inspection organization or by an external NDE inspection organization provided that the NDE procedures approved by NDE engineering are used by the personnel certified for fracture-critical NDE.*

c. NDE engineering shall be responsible for administering Special NDE certification tests and approving certification of the NDE personnel.

B.1.5 NDE Drawing Callouts

a. The hardware developer's responsible NDE engineering shall review the zoned drawings, the NDE inspection methods, procedures, and acceptance criteria.

b. Where there are different inspection requirements for different areas of a component, the drawing shall indicate separate inspection requirements for each zone.

B.1.6 NDE Process and Configuration Control

a. A written NDE procedure that complies with the relevant specification for the NDE method selected for the part is required for NDE inspection of each fracture-critical part.

b. Configuration control by revision or date shall be maintained for the personnel qualifications and certifications and for the NDE specifications, standards, and part-specific procedures.

c. Certifications shall remain current with revisions.

d. NDE engineering shall approve NDE process changes.

e. The RFCB and Technical Authority shall have final approval for NDE process changes that affect the reliability of fracture-critical NDE.

B.1.7 Capability Demonstration Specimens

a. NDE engineering shall be responsible for designing, approving, and providing Special NDE capability demonstration specimens.

b. If possible, NDE engineering shall borrow the specimens from NASA or other government departments.

c. If appropriate demonstration specimens are not available, the hardware developer or the sustaining engineering organization shall build or procure the demonstration specimens in concurrence with the RFCB and Technical Authority.

APPROVED FOR PUBLIC RELEASE – DISTRIBUTION IS UNLIMITED

d. Capability artifacts used in NDE procedure application shall be traceable to those used in the capability demonstration of the NDE procedure.

B.1.8 Responsibilities of NDE Inspection Organization

a. The NDE inspection organization shall be responsible for training, qualification, and facilitating NDE certification of its inspection personnel.

b. NDE inspection shall be responsible for maintaining and operating NDE equipment and facilities used in the NDE inspection of fracture-critical hardware.

c. NDE inspection shall be responsible for retention of certification records, NDE reports, and associated paperwork through the life of the program.

NASA-STD-5009A

APPENDIX C

REFERENCE DOCUMENTS

C.1 Purpose and/or Scope

This Appendix provides information of a general or explanatory nature but does not contain requirements.

C.2 Government Documents

	Koshti, A.M., SPIE 9437, "Ultrasonic measurement and monitoring of loads in bolts used in structural joints," Structural Health Monitoring and Inspection of Advanced Materials, Aerospace, and Civil Infrastructure 2015, 94370T (1 April 2015) (https://www.spiedigitallibrary.org/conference-proceedings-of-spie/9437/1/Ultrasonic-measurement-and-monitoring-of-loads-in-bolts-used-in/10.1117/12.2083220.short?SSO=1)
MIL-HDBK-1823	Nondestructive Evaluation System Reliability Assessment
SSP 30558	Fracture Control Requirements for Space Station
SSP 52005	Payload Flight Equipment Requirements and Guidelines for Safety-Critical Structures
NSTS 1700.7B	ISS Addendum, Safety Policy and Requirements for Payloads Using the International Space Station, Change No. 3, February 1, 2002
NASA/TM-2011-215869	A Comparison of the Capability of Sensitivity Level 3 and Sensitivity Level 4 Fluorescent Penetrants to Detect Fatigue Cracks in Various Metals, Parker, January 2011 (https://standards.nasa.gov)
NASA/TM-2014-218183	Interrelationships Between Receiver/Relative Operating Characteristics Display, Binomial, Logit, and Bayes' Rule Probability of Detection Methodologies, April 2014 (https://standards.nasa.gov)

C.3 Non-Government Documents

ASTM E746	Standard Practice for Determining Relative Image Quality Response of Industrial Radiographic Imaging Systems
ASTM E1647	Standard Practice for Determining Contrast Sensitivity in Radiology
ASTM E2002	Standard Practice for Determining Total Image Unsharpness and Basic Spatial Resolution in Radiography and Radioscopy
ASTM E2862	Standard Practice for Probability of Detection Analysis for Hit/Miss Data
ASTM E3023	Standard Practice for Probability of Detection Analysis for â Versus a Data
NTIAC-DB-97-02	Nondestructive Evaluation (NDE) Capabilities Data Book
NTIAC-TA-00-01	Probability of Detection (POD) for Nondestructive Evaluation (NDE)

APPROVED FOR PUBLIC RELEASE – DISTRIBUTION IS UNLIMITED

APPENDIX D

QUALIFICATION OF CR AND DR METHODS

D.1 Purpose and/or Scope

The purpose of this Appendix is to provide a table with CR and DR qualification methods.

Table 3—System Performance Tests for CR/DR

	Test	Method/Reference
Computed Radiology	SR_b (detector basic spatial resolution)	ASTM E2445/E2445M, Duplex wire ASTM E2002
	Plaque IQI sensitivity (visual) (contrast spatial resolution)	ASTM E1742
	Contrast sensitivity	ASTM E2445/E2445M, ASTM E1647
	Shading	ASTM E2445/E2445M
	Jitter	ASTM E2445/E2445M
	Banding	ASTM E2445/E2445M
	Erasure	ASTM E2445/E2445M
	Scanner slippage	ASTM E2445/E2445M
	Scan column dropout	ASTM E2445/E2445M
	Scan line integrity	ASTM E2445/E2445M
	Afterglow (Blooming/Flare)	ASTM E2445/E2445M
	Geometric distortion	ASTM E2445/E2445M
	EPS	ASTM E2445/E2445M, ASTM E746
	Geometric distortion	ASTM E2445/E2445M
	Photo multiplier tube non-linearity	ASTM E2445/E2445M
	Burn-In test	ASTM E2445/E2445M
	Spatial linearity	ASTM E2445/E2445M
	Central beam alignment	ASTM E2445/E2445M
	Imaging plate artifacts	ASTM E2445/E2445M
	Imaging plate response	ASTM E2445/E2445M
	Imaging plate fading	ASTM E2445/E2445M
	System spatial resolution	ASTM E2737
	Contrast sensitivity (uses CNR)	ASTM E2737

APPROVED FOR PUBLIC RELEASE – DISTRIBUTION IS UNLIMITED

Table 3—System Performance Tests for CR/DR

	Test	Method/Reference
	Signal-to-Noise ratio (SNR)	ASTM E2737
	Signal level	ASTM E2737
	Bad pixel mapping	ASTM E2737
	Gain calibration	ASTM E2737
	Offset calibration	ASTM E2737
	Plaque IQI sensitivity	ASTM E2737
	Lag Calculation	ASTM E2737
	Burn-In test	ASTM E2737
	Motion system repeatability	User selected
Display Monitors	Full modulation check	ASTM E2698
	Flicker check	ASTM E2698
	Distortion check	ASTM E2698
	1% line contrast	ASTM E2698
	5% line contrast	ASTM E2698
	Contrast ratio	ASTM E2698
	Luminous intensity	ASTM E2698

APPROVED FOR PUBLIC RELEASE – DISTRIBUTION IS UNLIMITED

www.ingramcontent.com/pod-product-compliance
Lightning Source LLC
Chambersburg PA
CBHW081638220526
45468CB00009B/2480